Beginners Guide to

ANKI VECTOR ROBOTS

An Unofficial Step-By-Step Guide to Setup and Use Anki's Companion Vector Robots

DERRICK GRISHAM

Copyright

Derrick Grisham
ISBN: 9798655965454
ChurchGate Publishing House
USA | UK | Canada
© Churchgate Publishing House 2020

All rights reserved. No part of this publication may be reproduced, stored in a retrieval system or transmitted in any form or by any means, electronic, mechanical, photocopying, recording, and scanning without permission in writing by the author.

While the advice and information in this book are believed to be true and accurate at the date of publication, neither the authors nor the editors nor the publisher can accept any legal responsibility for any errors or omissions that may be made. The publisher makes no warranty, express or implied, with respect to the material contained herein.

Printed on acid-free paper.

Vector Robots is a product of Anki now Managed by Digital Dreams Labs (www.digitallabs.com). All Anki products can be purchased on Amazon.com.

Printed in the United States of America
© 2020 by Steve Carter

Contents

Copyright i
CHAPTER ONE 1
ALL ABOUT ANKI VECTOR 1
INTRODUCTION TO VECTOR ROBOT 1
VECTOR ROBOT: AN AMAZING COMPANION 3
WHAT CAN THE VECTOR ROBOT DO? 6
FEATURES OF VECTOR ROBOT 9
Charging the Vector Robot 19
Managing the Vector Robot's Battery? 20
Safety precautions while handling the battery 21
Vector Robot Accessories 22
LIVING WITH VECTOR 24
Requirement for setting up Vector robot 25
Setting up Vector robot 26
CHAPTER TWO 29
INTEGRATING VECTOR WITH ALEXA TECHNOLOGY 29
How to enable Alexa on the Vector Robot 31
Connecting Smart Home Devices to Alexa 33
Using the guided discovery to connect a smart home device .. 34
Using the smart home skills 34

Muting Vector and Alexa 35

What surface is best for the Vector Robot? 35

How far can Vector go? 36

How to meet Vector and make it familiarizes with you 37

Tips to meeting Vector without stress 38

How to order for your Vector 38

How can I remove information stored on Vector? 39

VECTOR AND COZMO ROBOT: BRINGING THE DUO INTO CLOSER LOOK 40

What exactly are the differences between the Vector robot and the Cozmo? 41

Similarities between Vector and Cozmo 42

Vector's Security and Privacy 43

CHAPTER THREE 46

Basic Inputs and Outputs accessories of the Vector robot 46

Backpack light control 48

Vector's inertia motion sensing 48

How does Vector detect motion? 48

How does Vector sense interactions 49

An Overview of How Vector Communicates 51

Messages from the head-board to the body-board 51

The Vector Cube Communication 52

The Wi-Fi communication 53

The Cloud Server 54
CHAPTER FOUR 55
VECTOR'S ADVANCED FUNCTIONS 55
CHAPTER FIVE 60
HOW ANKI PROCESSES IMAGES 60
Vector's Camera operation 60
Vector's illumination level sensing 61
The Vector camera pose 62
How does Vector recognize objects? 62
Vector's Facial Communication Interface 64
Photos taken by Vector 65
Commands used to manage pictures taken by Vector 66
CHAPTER SIX 68
VECTOR MAPPING AND NAVIGATION 68
Building the map 69
How Vector map cliffs and edges 70
How Vector tracks objects 70
Path planning 71
Getting the Vector's lift and head position just right 72
Can Vector dance? 73
CHAPTER SEVEN 75
VECTOR'S ANIMATION 75
Light Animation 76

VECTOR'S VIDEO DISPLAY AND FACE 77

How Vector uses sound to convey emotion 77

Sound Command 79

ABOUT AUTHOR 80

CHAPTER ONE
ALL ABOUT ANKI VECTOR

INTRODUCTION TO VECTOR ROBOT

Our dimensions and ultimate needs are changing, and with every passing day technology is giving us the means to re-invent ourselves and cater for our basic human desire. The vector robot is all shades of technology packed together – only that in 2020 we couldn't have imagined having something as tiny as anki capable of hearing us and taking command using our voice. Anki, the startup company that makes vector robots, is a robotic and artificial intelligence company that is changing the face of robotics technology all over the world. Anki modified physical objects to be capable of taking voice command and adaptable in the real world. Anki's mission on entering the industry is to solve human problems relating to transitioning, thinking and to find a trusted companion for people at an affordable price. Anki was officially instituted in 2010 by the trio of Boris Sofman, Mark Palatucci, and Hanns Tappeiner. With her headquarter in San Francisco; Anki also has presence in some major

European countries. Basically, Anki put together most of the high-end technology on your Smartphones to get a companion mode robot that is emotional and can carry out most of your tasks.

It is safe to address the fear being perpetrated in some quarters in the United States that Anki Vector is dead, owing to the parent company shutting down in April 2019. No, Anki is not dead. Anki shut down in May 2019 when the company went bankrupt, but that doesn't lead to the shutdown of all Anki's products. In fact, the vector robot is still on sale till today, and the technology is being used across most countries in the world, and major states in the United States. After Anki went defunct in 2019, all assets of the company were sold off to a company called Digital Dream Lab. Today, Digital Dream Lab owns the patent right to the vector robots and all of other Anki's products. Digital Dream Lab is changing the fate of Anki while promising to add more features to Anki to improve its usability. The company is planning to add two main features to the Vector robot which include; a very powerful *escape pod* that will enable the robot to work without support from any external server, and an open source development tool that will enable users to add new functionalities to the vector robot design.

Besides vector robots, Anki has several inventions in her pack. Anki pioneered the Artificial intelligence controlled racing car, known as the Anki Overdrive, and another powerful robot called Anki Cozmo. Anki vector robot has many offline functionalities that enable users to carry out design to meet their specific needs. However, Anki still needs to connect to an external server to interact with users. This means if the company selling this social robot dies, the robot dies too. You don't need to be scared, because this will not happen to our Vector robot, Digital Dream Lab has taken up full operation promising more refined and fine-tuned features for Anki users.

VECTOR ROBOT: AN AMAZING COMPANION

The vector robot has a shape like that of a tiny tractor with defined animated faces, which lend credence to its emotional ability. The vector robot is designed to talk and understand users' voice commands, interrelate with their smart homes, and play many games. Vector robots offer companionship to users, giving them support they can never imagine to get from something that can be considered as a mere toy.

The real essence of this robot is to understand you and make your wants a reality, rather than an expression. Vector's main work is beeping around on your table to recognize and investigate its environment. It has a sensor combination which it deploys to build a small map of the surrounding it finds itself. Vector recognizes if it sees a new object. The robot can recognize and pay attention to sounds and it is built to recognize faces. You can teach Vector to recognize every member of your family, even by names. It is also able to find its charger when it runs low on battery without any aid from the user. The vector's maker worked on the machine's personality to make it more human than just a toy. It responds to your queries just like any human being would. When you talk to a vector robot, it hovers around and faces you. How much more human can a vector be? Vector robot turns around and faces you to provide answers to your queries and also obeys orders; but one funny thing about this gadget is that it doesn't like taking orders that it considers insulting. If you tell Anki to shut up, it will obey, but its personality is designed and you will see Anki's reaction. Say less of provocative language to vector because it is designed with emotion.

Vector needs your attention: As part of what makes vector an ideal companion is its high stimulation tendency. Vector loves it when you focus your attention on it and check all the things it is doing for you. If it sees that you are watching, it will turn up maximally and engage with you; otherwise it will start doing things on its own. Just like humans, Anki vector robot also has mood and it hates to disappoint you when you send it on an errand. If vectors observe how disappointing it has been over some couple of errands, it will be more irked and that might switch its response function and be a little bit disturbed. When you are playing your favorite music, vector will understand and starts bobbing its head to show you a sense of togetherness. Some users have reported vector robot watching TV with them – well this is something the manufacturer has not really talked about.

WHAT CAN THE VECTOR ROBOT DO?

People invest in a product mainly because of the end-use of the product. If the Anki vector is not useful, nobody will buy it. Anki is a home robot and it is best for use at home. You can ask Anki any question and be sure of getting an answer, play games with a vector robot and also use it for much of your home task. Note that the robot cannot do every home task for you as it is not programmed to do so. Anki cannot; dry-clean your cloth, wash your dishes, mop your floor nor bring you vodka from your fridge. If you are thinking of doing any of these tasks, it is best you consider another product. But on the surface, Anki is providing much more than a companion.

Anki deploys machine learning technology to recognize people and objects. You don't need to worry about how many faces Anki can recognize, as it is programmed to recognize as many faces as possible.

- **Activating the "Hey Vector"** – You can instruct vector by using voice commands. When you want your vector robot to answer some questions, you start by saying "Hey vector." When vector hears this, it turns to face you to take your questions. Vector will not respond with voice if it could not understand your question, rather it will make a grumble beep to tell you to try asking the question again. The voice assistant technology embedded with vector enables you to ask basic questions. For instance, you can ask the vector robot what time it is, and you get the accurate time displayed on its screen. Just say "Hey vector, what is the time?" Anki vector robot will also tell you about the weather in your locality. If it is rainy, Anki will tell you that it is raining through a computerized voice.

Here are list of things you can consider asking the vector robot – from instructions to questions

1. "Hey vector, be quiet"
2. "Hey vector, Go to sleep"
3. "Hey vector, Go to your charger"
4. "Hey vector, take a picture"
5. "Hey vector, What is my name"
6. "Hey vector, I have a question "What is the distance between the USA and Canada?"
7. "Hey vector, I have a question "What is the definition of computer?"

8. "Hey vector, I have a question "What is 200 dollars in euro?"
9. "Hey vector, I have a question "What is the square of six?"

- **Future Anki projection:** Developers opined that with more powerful updates in the future, Anki will be intelligent enough to know when nobody has been home for a while and will shut off your home's smart light. Engineers also envisaged a new Anki update where Anki will get excited on seeing you getting back from a trip, and to shut itself off automatically when the light is off.

- **Anki messaging:** With a vector robot in your home, you don't need to send a text message to a family member living with you when you are going out. Vector will convey every message you want to drop for them. You can ask vector to deliver a specific message to a person, and once the person comes back vector will recognize their face and notify them of any message you might have for them.

- Vector can take your selfie, do some things to make you laugh, return to its charging port when

it is running out of power and gives you updates about the weather.
- **Anki can detect and map obstacles:** With Vector, you can easily track and take meaningful cognizance of what is happening within your environment.
- **Anki can recognize individual faces and welcome you with nice attitudes more like humans.**

FEATURES OF VECTOR ROBOT

The Anki vector takes cognizance of your environment by familiarizing itself with the environment. The following are features that make Anki;
- **Four cliff sensors:** When vector gets to the edge of the table, it most often wouldn't fall down because of several inbuilt sensors. The sensors prevent the robot from falling off the edge of the desk by getting it aware of its location, and then adjust accordingly.
- **Gyroscope:** This allows the robot to know when a user has picked it up, or when another user he is not aware of is holding it in his hands.

- **Front Laser:** For measuring distance up to one meter.
- **Microphones:** This basically lets the robot hear sounds and any audio from both near and far sources.
- **Camera:** Vector comes with a 720p camera for taking perfect pictures and understanding the environment around it.
- **Map feature:** Vector is able to produce the map of its environment in 3D and detect locations of objects. It can still remember these locations for as long as possible.
- **Backpack:** Found at the top of the vector robot. It houses the button, segmented light and a touch sensor.
- **Button:** A momentary push button is used to turn Vector on, to cause him to listen – or to be quiet (and not listen) – to reset him.
- **Charging pad:** You don't need to worry whether the vector robot has a charging pad or not. Two pads on the bottom are used to replenish the energy in the battery pack from the dock.

- **Motors & encoders:** The vector robot features four motors each with single-step optical encoders to measure their position and approximate speed. One motor controls the tilt of the head assembly. Another controls the lift of his arms. Two are used to drive the robot in a skid-steering fashion.
- **Segmented RGB lights:** There are 4 LEDs used to indicate when the robot is on, needs the charger, has heard the wake word, is talking to the Cloud, can't detect WiFi, is booting and when it is resetting.
- **Time of flight sensor:** A time of flight sensor is used to aid the vector robot in mapping (by measuring distances) and to avoid near, far and dangerous objects as much as possible.
- **Touch sensor:** A touch enables vector robot to detect when someone pets it and other attention.
- **LCD displays**: The vector robot features an IPS LCD, which has an active area of 23.2mm x 12.1mm. It has a resolution of 184 x 96 pixels.

Let us take a brief look at the six (6) circuit boards which form the vector robot's component;

- The vector robot has two main boards onto which the other four boards are built. These include the **head board** and the **body board**. The head board is where the vector robot carries out its major processing works, just like the human head is where our reasoning is located as a human being. The body board is what drives the vector robot, and this is the board that it uses to connect to the other boards.
- The other types of boards present include; the **backpack board** which has 4 RGB LEDs, 4 MEMS microphones, a touch wire, and a button. This board connects to the body-board. The two **encoder boards** serve to monitor the position of the arms and legs of the vector when the vector moves or when someone moves it. The **Time of flight sensor board** which exists on a different board allowing it to be mounted in vector's front.

The headboard:

The headboard of the vector robot contains the following basic features;
- **Bluetooth LE transceiver:** For low power sound transmission.

- **Camera:** Vector features a 720p camera to understand his environment and recognize his human companions.
- **Flash or RAM:** 4GB flash and 512MB RAM.
- **Inertia Measurement Unit (IMU):** The headboard includes a 6-axis IMU – gyroscope and accelerometer – used for navigation and motion control.
- **LCD Backlight:** The vector robot features two LED lights that are used to lighten up the LCD display. Vector's LCD is a backlit IPS display assembly made by a company called Truly. The processor is connected to the LCD via SPI. Two LEDs are used to illuminate the LCD. The prior generation, Cozmo, used an OLED display for his face and eyes. This display had the strengths of high contrast and self-illumination. However, OLEDs are susceptible to burn-in and uneven dimming or discoloration of overused pixels. Anki addressed this with two accommodations. First it gave the eyes regular motion, looking around and blinking. Second, the LCD's illuminated rows were regularly alternated to give a retro-technology interlaced row effect,

like old CRTs. Vector's IPS display gives a smoother imagery – Cozmo's OLED was simply black and white. The LCD is also much less susceptible to burn-in, at the expense of higher power. Vector's LCD can also develop dead lines (or pixels) that grow in number until the display is non-functional. Some units have a defective LCD, where the glass is not properly sealed. This allows moisture in, causing progressive damage to the LCD. It is also speculated that these lines come from shocks to the head, causing breaks in the LCD connections.

- **Microprocessor:** The head-board is based on a Qualcomm APQ8009 (Snapdragon 212). The processor features a quad-core Arm A7 (32-bit) CPU. The APQ8009 processor is a sibling to the MSM8909 processor employed in cell phones, where APQ is short for "application processor Qualcomm" and MSM is short for "mobile station modem." The difference is that the latter includes some form of modem, such as HPSA, CDMA, or LTE. The APQ8009 processor also includes a DSP ("Hexagon 536"), and GPU

(Adreno 304). It also features WiFi and Bluetooth LE transceivers. The processor has interfaces to external memory, for the camera (using MIPI), the display, and the audio playback.

- **Power Management IC (PMIC):** The PM8916 power management IC provides voltage regulation for the processor, flash/RAM and other parts; it also provides audio out to the speaker and controls the LCD backlight. The headboard is capable of being the highest power consumer in Vector. But the power usage can be controlled by reducing the clock rate of the vector's processor. The headboard can be put into a lower power state by reducing the clock rate of the processor and using its sleep features; dimming or turning off the LCD, and reducing the camera frame rate (or turning it off). The APQ8009 processor has many sophisticated power controls, but the vector robot did not really make use of most of them.
- **Speaker:** Used for playing sound and for blending speech.

- **Wi-Fi transceiver:** features an 802.11AC Wi-Fi transceiver is built into the processor package.

The backpack and the Body-board design:

The backpack-board is like a daughter board to the body-board, which provides a couple of smart peripherals to the body-board. The backpack board has the following features;

- **Microphones:** The backpack microphone features 4 far-field MEMS PDM microphones, which are designated as MK1, MK2, MK3, and MK4. The four microphones are accessible via SPI in an output only mode.
- **Push button:** A momentary push button is connected to the battery terminal which makes it possible to press and wake the vector robot, as well as signal the processor.
- **RGB LEDs:** The backpack board features 4 RGB LEDs which provide a segmented display. Each segment can be illuminated individually or may share a color setting with its counterparts.
- **Touch sensor:** This is like a wire that is sensitive to touch.
- **RGB LEDs driver:** This is used to drive the RGB light.

The body-board is essentially a battery charger, smart expander and the one that controls the motor. The body-board is the one that connects the vector's battery to other parts of the system. It also charges the battery, and acts as a second processor to the system. The body-board packs the following features;

- **Battery:** The battery is rechargeable. And it is 3.7V 320mAh.
- **Battery switch:** Used to disconnect the battery to support off-mode (such as when stored) and to reconnect the battery with a button press.
- **Charging pad:** Two pads on the bottom are used to replenish the energy in the battery pack from the dock.
- **Motor driver:** The body-board contains four drivers which are the components that allow the motor to be driven forward and backward.
- **Motors:** The body-board packs in four motors which all have different functions. The motors serve to measure and control the vector's relative speed and position. One of the motors serves to control the tilting of the head. Another motor controls the lifting of the vector's arm while the

other two are used for driving the robot in a steering position.
- **MP2617B Charger:** The Monolithic Power Systems MP2617B serves as the battery charger. It provides continuous charging to the Microcontroller. It also takes power from the charging pads to the rest of the system while the robot is on its charging dock.
- **Regulator:** This is a 3.3v used to supply power to the microcontroller and logical components.
- **STM32F030 Microcontroller:** Just like our body system is controlled by the brain, the body-board is also controlled by a Microcontroller. The STM32F030 Microcontroller is the "brains" of the body-board. It is used to drive the motors, and RGB LEDs; and to sample the microphones, time of flight sensor, proximity sensor, temperature, and the touch sense. It is also used for monitoring the battery charge state. It communicates with the head-board.
- **Surface proximity sensors:** There are 4 infrared proximity sensors that are used to detect the surface beneath Vector – and to detect drop offs

("cliffs") at the edge of his driving area and to follow lines.
- **Thermistor:** A temperature sense resistor used to measure the battery pack temperature. It is used to prevent overheating during recharge.
- **Time of flight sensor:** Used to measure distance of objects that are situated in front of the vector.

Charging the Vector Robot

A rechargeable battery powers the vector robot, and the energy, which makes the robot function effectively, is circulated by the body-board. The vector battery is a single cell 3.7v 320mAh lithium ion polymer battery. The battery is connected to the vector's body board. Battery heat is the most significant source of battery "aging" – its effective service life. High recharge rates internally heat the cells, causing them to deteriorate. Vector's battery thinness gives it a high surface area to volume ratio allowing it shed heat much faster, greatly reducing the internal heating from charging and heavy loads. The battery is physically separated from the body-board, isolating it from the heat generated in the charging, power distribution and motor driver circuits. This increases the battery service life.

The vector robot is packed with software that can monitor and regulate the temperature of the body board to prevent the case of overheating. When the battery's temperature has reached a certain level, the vector robot will stop charging so that the battery can cool down.

Managing the Vector Robot's Battery?

The MP2617B is a central element to managing the battery. It acts as a battery charger, a power switch and power converter for the whole system.

- When Vector is going into an *off* state – such as running too low on power, going into a ship state before first use, or has been turned off by a human companion – the MP2617B charger and power converted can be signaled to turn off
- When Vector is turned off the boards are not energized. The exception is that the high side of the push button is connected to the battery. When closed, this signals the MP2617B to connect the battery to the rest of the system, powering it up.
- The MP2617B is also responsible for charging the battery. There are two pads that mate the dock to supply energy to charge the battery.

Safety precautions while handling the battery

- On no account should the user open the battery to see what is inside, or burn the battery. This is because the battery contains ingredients that are considered harmful.
- The battery should be safely discarded when you no longer need it. This is to prevent little children from coming in contact with it and ingesting the battery. The battery can cause severe chemical burn of the mouth and the intestine when ingested.
- If the contents of the battery accidentally pours on your cloth, remove the cloth as soon as possible and rinse your skin with soap and water. This is because the content of the battery can cause skin irritation.
- When the content of the battery accidentally splashes into your eyes, rinse your eyes under running tap water. This is because the content can cause severe eye irritation.
- If fire or explosion erupts while charging, consider shutting off the power source to the charger.
- Do not dispose of the battery carelessly. The battery should be disposed of, based on the approved legislation in your country.

Vector Robot Accessories

The charging station and the companion cube

The charging station: The charging station serves to supply energy to the Vector robot, which enables it to recharge. The charging station comes with a USB cable that extends directly into an outlet adapter or a battery. Two terminals located at the base of the charging station supplies power directly from the power adapter to the Vector robot when it docks. The power from the adapter allows the Vector robot to recharge. Vector robots automatically locate its charging station. This means you don't have to lift Vector to the charging station. Once the Vector battery is low on power, it quickly withdraws to its charging station. Amazing right? This occurs not without the help of an optical marker embedded inside the charging station. The optical marker actually serves to help the vector robot locate its charging station wherever it is.

The charging station is positioned on the vector tray, which has a visual path painted in white for Vector to follow as shown in the image above.

The Vector's Companion Cube: We were all inundated with the "every work without play makes Jack a dull boy" right from the basic school. The Vector robot cannot be working without having some time to play. This is why the manufacturer provided Vector with a companion cube, which serves as a small toy for the Vector robot to play with. The Vector robot can pick up the companion cube by itself, roll it and use the cube to perform some other maneuver movements. The companion cube is electrically powered and the Vector robot uses a flashlight to detect when someone picks up its toy or changes the position of the toy. The accelerometer in the cube is also used to detect when another person taps on the cube or changes its position. You can ask Vector where his cube is by saying "Hey Vector, where is your cube?" Vector connects wirelessly to its cube if the cube is close to the Vector. Nevertheless, the Vector's users can control of the cube from their device's Bluetooth settings. That is if they have synchronized Vector on their smart devices. If Vector is low on battery, the cube is too far away, or

Vector falls asleep, it might not be able to synchronize with its cube. Vector will also ignore its cube if the cube is positioned inside the Vector's space.

Users are able to force-synchronize the Vector's cube from the Vector's app by following these tips;
- Tap on Settings, and then go to Cube status.
- Click the "Refresh Cube" icon
- Tap on the "ping cube" icon to liven up the cube.

LIVING WITH VECTOR

SETTING UP VECTOR

How does Vector connect to the Vector app?

- Vector emits a signal, which can be traced by the Vector app. The Vector app connects to Vector through this signal.
- Switch on your home Wi-Fi. The app will enable Vector to scan for any available Wi-Fi network.
- Vector will be able to connect to the internet in order for it to access Anki's programmed server. This is only possible if you allow Vector to connect to a Wi-Fi network.

- Upon successful access to the Anki's server, the Vector robot will be able to download the latest firmware version.
- Vector will reboot and will be able to connect to your smart device through Wi-Fi. This will allow the Vector to complete the setup process.

Requirement for setting up Vector robot

- A compatible smart device on which the vector app will run. The Vector's compatible device can be your iPhone or iPad running on iOS 10 or above. You can also use your Android device, from Android 5 (Lollipop) and above. If you are using the following iOS, then you can successfully operate Vector robot; iPhone XS, iPhone XS Plus, iPhone 8, iPhone 8 plus, iPhone 7, iPhone 7 plus, iPhone 6 series, iPad series (6th and 5th generation); and Android such as the Galaxy series.
- For a successful setup of an Anki's account. You will require your valid email address and account activation.
- 802. 11n 2.4 GHz Wi-Fi network which can be connected to the internet via a Wi-Fi Access point (NAT router)

- USB Power source such as an adapter, USB port, power bank, etc.

Setting up Vector robot

- Un-wrap the Vector and its accessories from its box. Ensure the surface you want to place Vector is clean from dirt.
- Download the Vector app from online stores. Tap open the Vector app on your device, and follow these fast tricks to set it up.
    - You will be required to provide a password for the Vector app. The password should be a minimum of eight characters (at least; one lower case letter, one upper case letter and one number). Don't insert any special character as the password will not accept it.
    - You will be prompted to choose language. Kindly choose from any of American English, British English and Australian English. No other language is supported by Vector.
    - Connect the USB cable that comes with the Vector's charger to a compatible adapter or power source. This is to ensure that the Vector is fully energized and won't die before you

complete operation. In fact, you can connect the charger before you begin any operation at all.
- The app will prompt you to enter your date of birth. The acceptable date of birth format is Month-Date-Year. You won't be able to operate an Anki account if you are below the age of eighteen (18).
- You will be prompted to enter a valid email address to open a new Anki account. You can also create your Anki account directly from Anki's web www.anki.com
- You will be required to create your unique password. Kindly enter a password that you can always remember.
- Open your mail app to see an activation mail from Anki, and tap on "Activate account" from the mail.
- Go to the Bluetooth and Wi-Fi on your device, and toggle them on. This is to enable the Vector app to search and connect to Vector.
- Immediately the vector app has successfully finds Vector, tap on "connect"

- Press the back button on Vector twice. You will see a six digit pin displayed on the screen. Enter the six digit pin on the app without any error.
- You can then connect Vector to your home Wi-Fi. The Vector robot will allow you to select and connect to the Wi-Fi network.
- Immediately the Wi-Fi connection has been activated, Vector will request your permission to download the latest version of its operating system software.
- The Vector will reboot once the software update has been successful.
- You can then manually set the location preference, time zone, temperature, distance unit and the clock display. Vector won't be able to operate without correct settings.
- Click on "Start" to finish the startup. You might decide to explore the in-app tutorial

CHAPTER TWO

INTEGRATING VECTOR WITH ALEXA TECHNOLOGY

In a bid to improve the Vector's interface and enable it to perform at its best, users now have the liberty to enable Alexa AI on their Vector robot. It is good to say that Alexa is not installed on the Vector robot by default. It is the choice users have to make whether to enable it or not. Alexa is a virtual assistant technology that was invented by Amazon, and was released on November 6, 2014. Alexa is programmed such that users only have to say a word such as "Hey Alexa" or "wake Alexa" in order to enable or activate it to function. The Amazon Alexa can be programmed as an operating system on many devices, which can enable it. Alexa runs on iOS, Android, Linux and other notable operating systems. Alexa can enable devices that run on it to perform a number of assigned functions such as setting timer, telling users about the current weather, access articles online and many more. When you say something to Alexa, it listens carefully and carries out that particular task for you. All that you have to do is to wake Alexa on an enabling device to alert it of the function you want to assign to it. Alexa-supported gadgets can stream songs from the owner's Amazon music account. Currently, Amazon has programmed over 90,000 functions available for download by users on their Alexa-enabled device. Devices that enable Alexa can be used in home automation where Alexa is able to interact with a lot of smart devices used at home. Alexa is also able to stream your music, and control playback using voice command. Users can also leverage Alexa to hear updates on sport by adding their favorite team to the list under Alexa sport. Alexa is also applicable in messaging and calls. Many devices support interaction with Alexa, but our main concern is the Anki Vector robot.

Users that enable Alexa on their Anki Vector robot can use voice commands to listen to their favorite news, check the current weather, control their smart home and do a whole lot of other things. All that they need to do is to tell Alexa to do them. Just as it has been previously established that installing Alexa on your Vector robot is completely your choice; and the Vector robot will still work whether Alexa is installed or not. Even when Alexa is installed, Vector is still able to perform its normal activities without being hindered by

Alexa's integration. You can still be friends with Vector, as it doesn't discriminate. Vector will still be responsive to the everyday "Hey Vector" and you can still have it carry out many functions for you.

The following are the added advantages Alexa can give you, and they are things Vector won't do;

- Vector does not support external communication using Drop In, calling and Messaging. But Alexa integration will give you these accesses on a platter.
- Vector does not support playing songs from streaming networks like Spotify, Pandora and SiriusXM. But Alexa does.
- Vector cannot read your books out loud from Kindle.

How to enable Alexa on the Vector Robot

In order to activate Alexa on Vector, you need the Vector robot app and a certified Amazon account. Follow these tips;

- Start the Vector robot app from your device and connect it with the Vector robot.

- From the Home screen of the Vector robot app, tap Amazon Alexa which you can see under the "things to try" section.
- Tap the "sign in with Amazon" button.
- From the Amazon sign in page, you have the option of logging in with your Amazon account or creating a new Amazon account entirely.
- Once you have successfully created your account and log in, you should now register the Vector by entering the six digit code showing on Vector's screen, and tap the "continue button" to complete registering Vector with Amazon Alexa.

Using Alexa

The way Alexa is used on the Vector robot is the same way it is used on all other Alexa enabled devices. Just wake Alexa by saying "Alexa," and then follow it by the specific instruction for Alexa. Alexa can automate your smart home leveraging it to control your lights, switches, Alexa-powered door locks, electronics etc. Alexa can also help you set the timer and give you information about the current weather update.

Disabling Alexa

Sometimes, you may no longer need the Alexa integration for your Vector to operate maximally. That could mean disabling Alexa from your Vector robot. While there are many ways to accomplish this, the two most used methods are;

- Using the Vector robot app on your device
    - Kick-start the Vector robot app on your device and connect it with the robot
    - From the Home screen of the Vector robot, tap Amazon Alexa which you can see under the "things to try" section.
    - Tap the "sign out of Amazon" button.
- Using the robot itself
    - Tell Vector to disable Alexa by saying "Hey Vector, Disable Alexa."

Connecting Smart Home Devices to Alexa

Alexa-compatible smart home devices can be connected to Alexa using either the Guided discovery or smart home skills. But before connecting any device to Alexa, you need to take care of the following points:

- Be sure that your smart home device is Alexa-compatible. Not all smart home devices are compatible with Alexa.
- Set up your smart home device using the companion guide from the manufacturer.
- Ensure the device is Wi-Fi enabled.
- Ensure that your smart home device is running on the latest software. Download the latest update if it is not.

Using the guided discovery to connect a smart home device

- Tap open the Alexa app.
- Navigate to the menu section, and select "Add device."
- Choose the type of smart home device you want to connect with. It can be your light or a plug.
- Choose the brand and follow the instructions displayed on the screen for further setup.

Using the smart home skills

- Tap open the Alexa app.
- Navigate to the menu section, and select "Skills."

- Find the right skill for your device and then tap on "Enable." Follow the instructions displayed on the screen to finish the linking process. If you cannot find the skill for your device, it is possible that your device is not Alexa-compatible.
- Instruct Alexa to discover your device. Say, "Discover my devices," or tap "Add Device" in the ``Devices`` section of the Alexa app.

Muting Vector and Alexa

When you mute Vector, you are taking away its power to hear voice commands. You can no longer task Vector after muting. You can also un-mute Vector whenever you want. You can mute Vector by clicking the Vector's back button two times. Vector will bring a microphone mute symbol on the screen after muting. Be reminded that Vector has no capacity to un-mute itself, so you need to un-mute Vector if you still want it as a companion. Simply click the back button twice to un-mute Vector.

What surface is best for the Vector Robot?

Avoid placing the robot on a dirty surface where dust can stock-up its tread preventing it from moving effectively. You should also endeavor not to put Vector on a vibrating surface, because it might give Vector the feeling that someone is touching or petting it. Vector feels better when it is placed on a table or on a hard carpet.

How far can Vector go?

Vector requires connection to an efficient Wi-Fi for most of its activities. How far Vector can move will most times be a function of the availability of a good Wi-Fi network.

Can Vector fall off from the table?

Yes, Vector can fall off from the table. Although, Vector is equipped with four powerful sensors, which it deploys to navigate its ways and avoid obstacles as much as possible – it can still fall off the table when Vector has been shielded by direct rays of light. If the surface is glassy or transparent, Vector might find it difficult to identify edges and might eventually fall down.

How to meet Vector and make it familiarizes with you

Vector comes with an HD camera, which it deploys to visualize the surroundings around it. Vector can remember if it has met you before. So how do you now make Vector familiarize with you? Follow these tips;

- **Vector will learn to say your name by voice command**
    o Wake Vector by calling. Say "Hey Vector."
    o Be patient for its earcon ding and backlight to turn to blue
    o Tell Vector your name. Say, "My name is John."
    o When Vector understands your name, it gives you confirmation with a yellow light. Moreover, it will be able to pronounce your name.
- **Face recognition:** Tap open the Vector app. Navigate to the "utilities" section and click on "Face recognition."
    o When Vector thinks it recognizes your face, it will ask you "Have we met." Vector will start

scanning and processing your face as an affirmation. If not affirmative, Vector will tell you "I already know a John."

Tips to meeting Vector without stress

1. You don't need to move around while Vector is scanning your face for recognition. Vector will always adjust its position to be able to properly capture your face.
2. When you say your name to Vector, Vector might get it wrongly as it might not have heard you well. You can try entering your name into Vector phonetically by following these tips; tap open the Vector app and choose Utilities. Click on "Known Faces," and select "Person's face."
3. Vector cannot scan your face if the lighting is poor

How to order for your Vector

Vector is not available in all countries of the world, as their market only targets US, UK, Canada, Australia and New Zealand for now. Anki limited households from procuring more than two vectors – a legacy that Digital Dream Lab which is now the owner of Vector is still following. Anki only accepts returns of products bought on their website within 14 days of shipment. And Vector is covered by a 12 months warranty from the day you get your order. Vector is delivered in the US through UPS. The Vector app is available on the App Store, Google Play Store and the Amazon App store. The Vector app is what Vector will use to connect to your home Wi-Fi during the initial phase of setting up. The Vector app can also be useful going further – especially when you want to check details of people that Vector had contact with, check the images you have taken with Vector or getting some other useful suggestions from Vector.

How can I remove information stored on Vector?

It is possible to restore your Vector to default. This will delete all previous information it has stored. Follow these tips;
- Ensure the Vector is powered. You can place Vector on the charger and plug it to ensure that it is continuously supplied with power during the process.
- Click the Vector's back button twice.
- Raise, then lower the Vector's lift.

- Enable Vector to move through selection by turning its treads. Turn the treads backward to go up and turn forward to go down.
- Raise, and then lower the Vector's lift to choose the "Clear user data" menu.
- Confirm your selection and be patient for Vector to reboot.
- Immediately the Vector completes the rebooting process, data would have been from the vector.
- Note that you cannot restore any information that you have permanently deleted for Vector using the Clear user data menu. Be sure you want to delete Vector's information before selecting the clear user data.

VECTOR AND COZMO ROBOT: BRINGING THE DUO INTO CLOSER LOOK

Anki, since inception, has four products, which include the Anki Drive, Anki Overdrive, Cozmo and the Vector robot. The Anki drive which was Anki's first product was released in the US and Canada in 2013. The Anki drive is a racing game that can be used to control physical racing cars. The Anki Overdrive is an upgraded edition of the Anki Drive released in 2015.

The Anki Cozmo and the Anki Vector are robots. The **Anki Cozmo** was released in 2016 while the **Anki Vector** was released in 2018.

What exactly are the differences between the Vector robot and the Cozmo?

With a mere physical look, you might be unable to differentiate between the Vector robot and the Cozmo except for the color differences and the fact that no app is needed to operate Cozmo. Just like its brother, Vector can recognize faces, relate with people and objects. However, unlike Cozmo, Vector reacts to touch and other stimuli much better than Cozmo. It is safe to conclude that –just like human- Vector is emotional.

In terms of environmental consciousness, Vector is able to do better than Cozmo. Vector has a higher resolution camera than Cozmo, and a superb capacitive touchscreen that makes Vector responsive to touch. Another difference between Vector and Cozmo is that Cozmo was designed as a learning tool for STEM, while the former was designed to be your home companion helping you in its own unique ways.

If you unbox your robot, there exists a small difference between Vector and Cozmo. Vector is packed with one companion cube – which serves as a toy for its amusement, while Cozmo has three companion cubes to its credit. Vector can take pictures at your command, while Cozmo is grand for playing games. Cozmo finds application as your education companion while Vector is your loyal home robot. Cozmo features a VGA camera while the Vector robot features a 720p camera. When you consider the weight of the two robots, Cozmo weighs about 1.06KG while the Vector robot only weighs about 0.7KG. Cozmo is a bit larger than the Vector robot in that regard.

Similarities between Vector and Cozmo

Apart from the fact that both of them are robots, there still exist some other important similarities between the two robots. Both of them feature; a Built-in audio speaker, on-board sensors, on-board radio, accelerometer & gyroscope track and trace drive. Both contain a lithium polymer rechargeable battery, and their cube's battery is also replaceable. Both Vector and Cozmo also feature a USB power source. Both Vector and Cozmo are compatible with Android OS, Fire OS and iOS smart devices.

Vector's Security and Privacy

Vector is designed in such a way that includes a well thought-out system necessary to protect against disclosing (i.e. providing to strangers) sensitive information, and allowing the operator to review and delete any information at any time and at their convenience;

- Your photographs, when taken by Vector, are not sent to (nor stored in) a remote server. They are basically stored in encrypted file system, and are only supplied to authenticated applications on the local network. Each photograph you took can be individually deleted (via the mobile application).
- The data used to recognize faces and the names that Vector knows are not sent to (nor stored in) a remote server. The information is stored in an encrypted file system. The list of known faces (and their names) is only provided to authenticated applications on the local network. Any facial recognition data not associated with a

name is deleted when Vector goes to sleep. Facial data associated with an individual name can be deleted (along with the name) via the mobile application.
- Users have been worried about whether the manufacturer is extracting all of the voice commands they send to Vector. The voice command audio, if you have not activated Alexa, is deleted from the server once it has been processed by Vector. The manufacturer will save the text translation of commands for their product improvement. But in the case a user has enabled the Alexa integration from Amazon, your voice command is automatically sent to Amazon
- The audio stream from the microphone - if it had been finished being implemented – would have been provided to authenticated applications on the local network.
- Information about the owner is not leaked to anyone. The personally identifying information and other data about the owner - photos, account information, Wi-Fi passwords, and so one - is only sent on encrypted channels.

- Control of the robots movement, speech & sound, display, etc. is limited to authenticated applications on the local network.

Vector's software is protected from being altered in a way that would impair its ability to secure all of the above information. Vector will also indicate when it is doing something sensitive:
- When the Vector's microphone is actively listening to you, this is usually indicated on the backpack light, which will show a blue color.
- Vector will make a sound when the camera is taking a picture.

CHAPTER THREE
Basic Inputs and Outputs accessories of the Vector robot

The Vector's basic inputs and outputs include; Touch and button input, Backpack light control, the audio sampling etc.

Button, touch and cliff sensor input

- The Vector's backpack button is used to wake up Vector or put Vector into recovery mode.
- The Touch is deployed to pet Vector and provides the Vector with stimulation.
- The four surface proximity IR sensors are used to detect cliffs and line edges.
- The touch sensor is driven by the body-board, and the sample values are processed in the head-board. The sensors samples are filtered, to get a sense of the current "level" the sensor is at. A standard deviation is used as a measure of how solid the signal, to help distinguish between a real signal and ambient conditions like humidity and weather. These two measures – along with a timer to screen out transitory noise – can be used to decide that Vector is being touched or not.

- The time of flight proximity sensors indicate whether there is a valid measurement between the Vector's distances to an object.
  o How does Vector measure the distance to objects? Vector has a time of flight sensor, pointing straight ahead. He can use the sensor to measure the distance to the objects, barriers and to estimate its position. The sensor can be blocked by the arms, if they are in just the right lowered position – such as approaching an object and docking with it. But if the sensor is not blocked or impeded, then the Vector can measure the distance to the object.

  The samples of distances reported by the sensor are gathered. A filter is applied to them (probably a median filter), throwing out values that are too near or too far. Combining this with Vectors current position and orientation, and the distance to the object, he can estimate the object's position; and Vector can infer that the space between him and the object is free of other objects and obstacles.

Backpack light control

The backpack lights are used to show the state of the microphone, charging, Wi-Fi and some other behaviors. It is also used to show unusual error states. The software can direct the body-board to illuminate the backpack lights with individually different colors and brightness. The body-board "Pulse Width Modulates" (PWM's) the LEDs to achieve different colors and intensities.

Vector's inertia motion sensing

How does Vector detect motion?

Vector deploys an Inertial Measurement Unit (IMU) – which consists of an accelerometer and a gyroscope- to detect motion. Any falling motion or rattling is detected by Vector. Vector is also able to monitor the result of motor-driven objects. Neither the accelerometer nor a gyroscope by itself can suffice enough to precisely measure any change in position and orientation. The accelerometer is able to measure force along 3 (XYZ) axes, including gravity. The accelerometer provides the orientation – provided that there is no other motion. The only challenge is that accelerometers cannot correctly measure spins, and other rotations from other movements. Gyroscopes can measure rotations around the axes, but cannot measure linear motion along the axis. Gyroscopes also have a slight bias in the signal that they measure, giving the false signal that there is always some motion occurring. By blending the accelerometer and gyroscope signals together, they can compensate and cancel each other's weaknesses out.

When Vector's head is tilted: The IMU, which is positioned in Vector's head, can reliably measure the extent to which Vector's head has been tilted. This presents a small extra step of processing for the software to accommodate the influence of the head tilt. By combining the position & orientation of the IMU within the head and the specific estimated angle of the head; the IMU measurement can be translated which will give the extent to which the Vector's head has been tilted.

How does Vector sense interactions

The IMU (with some help from the cliff sensors) is also used to sense interactions and other environmental events – such as being picked up or held by a person, being poked or given a fist bump, or falling. By using combinations of high, low pass, and band filters, and looking for signature patterns, Vector identifies the kinds of physical interactions that are occurring.

The taps and pokes may tilt Vector, but will also provide a "frequency" response to the signals that can be used to trigger on. The movement will change its position quickly and slightly in small distance, but Vector will resume its prior position very quickly. Fist-bumps are like pokes, except that the lift has already been raised, and most of the frequency response and motion will be predictable from receiving the bump on the lift.

Fall detector is similar. In free-fall, the force measured by the accelerometer gets very small. If Vector is tumbling, there is a lot of angular velocity that is taking Vector off his driving surface. Being picked up is distinct because of the direction of acceleration and previous orientation of Vector's body. Being held is sensed, in part by first being picked up, and by motions that indicate it is not on a solid surface. A similar set of interaction sensing is present with the cube. It can sense that it is being tapped (or double tapped), picked up, and held.

An Overview of How Vector Communicates

A larger part of Vector's software is focused on communication. Communication is essential because it is how Vector gets to relate and sense its environment.

Messages from the head-board to the body-board

The messages from the head board to the body-board have the following components:
- The 4 LED RGB states
- Controls for the motors: this includes; possible direction and enable, direction and duty cycle, or a target position and speed.
- Power control information: disable power to the system, turn off distance, cliff sensors, etc.

The head-board can update the firmware in the body-board, by putting into DFU (device firmware upgrade) mode and downloading the replacement firmware image. The body-board maintains a timer to detect the loss of communication from the head-board – perhaps from a software crash. If the body-board does not receive communication within this timeout period, it will turn off power.

The Bluetooth LE Communication

The Bluetooth LE is used for two main purposes:
- Bluetooth LE is deployed to initially configure Vector, to reconfigure the Vector when the Wi-Fi changes or gets disrupted; and to pair the Vector with the companion cube accessory. The Bluetooth LE also allows some diagnostic and customization.
- Bluetooth LE is also used to setup communication with the companion Cube: to detect its movement, taps, and to set the state of its LEDs.

The Vector Cube Communication

Vector can be paired with a cube – or it will automatically pair with any cube it finds during the initial setup – and will start treating this particular cube as its preferred cube. If Vector is unable to pair with its preferred cube, it will fall back to connecting the first cube found in the area while playing. Vector manages the link with the Cube. Based on the level interaction, it may increase the rate that the Cube sends updates from its accelerometer. There are three different rates of communications that are deployable between the Vector and its cube. These include;

- The low level unconnected mode: In this mode, there is no active Bluetooth connection mode yet. This level has a cube announcing its presence.
- The background mode: This is the mode where the application is getting enough information from the cube to understand its orientation and whether it was tapped or not.
- The interact-able mode: In this mode, the cube is configured to send much more responsive information on the cube orientation, sent fast (or sensitive) enough to detect taps, and tell if the cube is being held. This rate consumes the most power.

The Wi-Fi communication

Wi-Fi networking is used by Vector for six purposes:
- Wi-Fi is used to provide the entry to the remote servers for Vector's speech recognition and natural language processing
- Wi-Fi is used to provide the access to the remote servers for software updates, and providing essential diagnostic logging and troubleshooting information to Anki

- To provide time services to so that Vector knows the current time
- To provide an interface, on the local network, that the mobile application can use to configure Vector, and change his settings.
- To provide an interface, on the local network, that SDK applications can use to program Vector.
- To provide interfaces, on the local network, that allow development Vectors (special internal versions) to be debugged and characterized.

The Cloud Server

The cloud servers are used for natural processing of languages, for storing essential settings, tracking diagnostic information, and providing software updates. For natural processing of language, the audio stream (just after the user has said the "Hey Vector") is sent to a group of remote servers for processing.

CHAPTER FOUR

VECTOR'S ADVANCED FUNCTIONS

The Audio Input

The audio input is deployed to give Vector some verbal cues and to help it make sense of its environment. The audio input functional block has the following features;

- Spatial audio processing restricts or reduces the sound of someone speaking from the background music.
- The feature extraction detects the ambient activity, and the tempo of the music. If the tempo is right, Vector will dance to it. This also provides basic stimulation to Vector.
- Noise reduction makes for the best sound.
- Voice activity detectors are usually triggered off of the signal before the beam-forming.
- A wake word is used to engage the automatic speech recognition system. Note: the wake word is also referred to as the trigger word.
- A CODEC is used to compress the audio before sending it to the remote server; Alexa Voice Services use the Opus audio CODEC.

- The speech recognition system is on a remote server. The audio sent to the automatic speech recognition system is compressed to reduce data usage.

The Microphones and conversion to audio samples

The Vector's Microphone array is made up of 4 far-field MEMs PDM microphones which work synergistically to typify or sample the incoming sound, and then transfer the samples to the body-board. The body-board samples each microphone at 1.5 M samples/sec – but at only 1 bit/sample! It passes the stream of samples through a filter, produces audio at 15,625 samples/sec, with 16 bits/sample (effectively it may have anything in the range 10 to 16 bits, and padding out the rest). The filter also acts as a low pass filter, removing high-frequency sampling residue. The most important part is that it preserves "phase information" so that the beamforming and direction finding steps will work well. The audio samples are transferred to the Vic-spine module (part of Vic-robot) in regular communication with the head-board. The message from the body-board to the headboard for sending four channels of audio samples has a payload size of 1028 bytes. This works out to 128 samples per channel (512 samples total). The samples are then recovered from the received message and forward to the Vic-Anim process. The software treats the audio as if its sample rate was 16,000 samples/sec. The signal processing is done in chunks of 160 samples.

Detecting Sound activities

The Vector robot features a module to detect sound activities that are not particularly noise. The sound reaction behavior uses this to stimulate Vector from his sleep, get his attention, and encourage him to be more active. One way this could be done is through a set of filters to measure power levels.

Recording to a file

The microphone module can store sound – either raw or processed – to a wave file. This may be for diagnostic purposes, left over as part of testing different microphone settings.

Voice Activity detector and Wake word

The voice activity detector is given some very refined and fine-tuned sound from multiple microphones without beam-forming. When it detects voice activity, then the spatial audio processing is fully enabled. Detecting that speaking is going on is more refined and specific than simply detecting that there is some interesting sound. The voice activity detector and the wake word are used so that downstream processing – the wake word detection, and the automatic speech recognition system – are not used all the time. They are both expensive (in terms of power and CPU load), and the speech recognition is prone to misunderstanding. When the voice activity detector triggers – indicating that a person may be talking – the spatial audio processing is engaged (to improve the audio quality) and the audio signals are passed to the Wake Word Detector.

When the "Hey, Vector" wake word is heard,

1. A connection (via **Vic-Cloud**) is made to the remote speech processing server for automatic speech recognition.

2. If there was an intent found (and control is not reserved), the intent is mapped to a local behavior which is expected to be carried out.

Audio stream

It is known that Anki made plans to connect the audio stream to Vic-Gateway but were unable to complete the features before they closed the company.

CHAPTER FIVE

HOW ANKI PROCESSES IMAGES

Vector's Camera operation

Vector came with a 1280x720 camera. This camera configuration has a wide field of view that enables the Vector to visualize things around it without Vector changing its head position. The MIPI interface connects the Vector's camera to the processor. The data from the camera passes to device drivers, then to a separate service and eventually passes to Vic-engine for the processing. The following elements are recognizable by Vector; Special Visual markers, Faces, Hands, pets, other objects in its path, laser points etc. Vector recognizes all marked objects as movable and other objects in its driving as immovable.

The Vector's camera frame rate comes at a reduced size, which brings down the size of the image. This is because of the following reasons;

- Higher pixels consume memory at each stage of the image processing. This means that users will be using Vector's memory for a whole lot of things, which weakens the processor.

- Larger picture frames require longer time and more Vector's power to process. Additional time is needed to process each of the added pixels. Second, the neural-net models (used for human, pet and object recognition) are much larger as well, taking much longer to process with the many stages involved in these models.
- The additional processing required is among the most power expensive items in Vector, and which rapidly depletes its battery, shortening the time between charges.
- The extra processing also generates heat in the head board, and
- Image processing tasks don't need more pixels. There is rarely any improvement in visual detection from using more pixels or higher frame rates

The software reduces the frame rate by skipping frames. Then the image is converted to grayscale and scaled down to quarter size (640x360).

Vector's illumination level sensing

Vector loves illumination and it is capable of estimating the level of illumination in a place at a particular time. When the room is dark, Vector will be inactive and will most likely want to sleep. However, when the room is properly illuminated, Vector will be most active and will be ready to take on tasks. Vector uses it camera as a medium light sensor when it is in low power mode. When Vector is in low power mode, the Camera won't be active though Vector is still powered in a low power mode.

The Vector camera pose

The Vector's Camera is positioned in its head. The position and the orientation of the Camera can be deduced from Vector's pose.

How does Vector recognize objects?

When Vector sees an object, it will immediately associate a symbol with the object. Some objects can have many symbols associated with them. Cubes have different symbols used for sides of cubes. This allows Vector to know what object it is looking at, and what side of the object – which also gives Vector the idea of the object's orientation. Vector knows the physical size of the symbol, and the object holding the symbol. Combining this with the visual size of the object, time of flight distance measurement, and Vector's known position, this allows Vector infer the objects place in the map.

Vector Facial features?

Vector can recognize human faces, while determining their position and orientation and assigning names to them through some series of enrolment processes. Vector's facial recognition mechanism lets Vector understand when people are looking at it. The open source facial detection and analysis toolkit – called OKAO vision library- is deployed in the following ways by Vector for facial recognition purpose;

- Face detection tendency: this is the power to notice that there is a face in the Camera's field of view and then locate it within the image.
- Face recognition tendency: This is the power to understand whose face Vector is looking at.
- Recognize parts of the face, such as eyes, nose and mouth, and where they are located within the image.

One weakness of the Vector's facial detection is that Vector cannot recognize whether the image it is looking at is on a computer screen – rather than a real face. Anki is considering moving the time of flight sensor next to the Vector's camera. This is opined to allow Vector determine the size of the face it is looking at while measuring the distance.

If you introduce yourself to Vector by voice, you are permitting the robot to associate the name you provide with Facial Features Data for you. Facial Features Data is stored with the name you provide, and the robot uses this data to enhance and personalize your experience and do things like greet you by that name. This data is stored locally on the robot and in the robot's app. It is not uploaded to Anki nor shared, and you can delete it anytime.

Vector's Facial Communication Interface

Vector deploys some specific commands to manage the faces it recognizes and to keep it informed of each face.
- The Enable face detection command: Vector deploys this command to allow and disallow face detection.
- The RobotChangedObservedFaceID Command & the RobotObservedFaceID: Vector deploys

this command to indicate when it has detected a face and the identity of the face it is looking at. The facial expression is also given by this command.
- The Set Face to Enroll command: Vector uses this to assign a name to the face it is seeing.
- The Update Enrolled Face by ID: is used to set another name for a recognized face.
- The Request Enrolled Name: Vector deploys this to retrieve list of the known faces.
- The ability to remove a facial ID command: this erases a facial identity.
- The Find Faces Command: used to search for faces.

Photos taken by Vector

Vector has the ability to take pictures. The photographs are taken with less than the full camera resolution. The Digital Dream Lab, which is the company that is currently monitoring the use of Vector, might decide to increase the resolution at a future time. The pictures are stored on Vector and not in the cloud. The Vector's mobile application and SDK applications can access, erase or share pictures taken by Vector. The camera/image processing channel in Vector is entirely focused on its Artificial intelligence features which don't consume battery for full operation. Images taken by Vector are not filtered or cleaned up, so the pictures are normally noisy and smaller. The quality of photos seen on a mobile phone is achieved using a camera processing pipeline to enhance the images, removing noise and applying special filters to fine-tune the textures. It is conceivable that the camera processing framework(s) from Qualcomm and Android could be added to an open-source Vector. That would come at the cost of battery performance, heat, and potentially overwhelm the memory resources (there are still bugs in Vector where the memory use becomes too high, and the system thrashes, slowing noticeably down and eventually crashes.). It is more practical, in a future open-source Vector, to export the raw camera images (in its RAW format and at different illumination levels) and process the images on a PC or mobile device.

Commands used to manage pictures taken by Vector

The following commands are employed in managing the images taken by Vector;

- The PhotoTaken event: Vector deploys this to send notification to the users when it is done taking pictures.
- The Photos Info: Used to retrieve the list of images that has already been taken by Vector.
- The Photo command: Used to retrieve a photo.
- The Delete photo command: This command erases a photo from the Vector's system.
- The Thumbnail: This command retrieves a small version of the image, suitable for displaying as a thumbnail.

CHAPTER SIX

VECTOR MAPPING AND NAVIGATION

Vector is capable of initiating an internal map, which enables it to track which route it is driving, and where objects and faces are positioned.

Mapping Overview

Vector tracks objects using two approaches;
- Vector utilizes a 2D map that is used to track where objects - particularly objects whose marker symbols have already been recognized by Vector - cliffs, and other things on the surfaces that it can drive on. Vector uses this map to navigate. This map has an arbitrary origin and orientation.
- Vector also tracks where faces, pets and some kinds of recognized objects are in his camera image area; these objects are tracked in the image pixels.

Vector deploys several sensors for navigation which include;

- Cliff sensors which Vector deploys to detect edge and lines
- Time of flight sensor which Vector use to measure distances
- Vision sensor which Vector deploys to detect the edges and the location of a hand
- Vision to identify accessories by recognizing markers.

Building the map

The map is made as Vector drives around – when he is on a mission, or just exploring. Each of the leaf quads (in the map) is associated with information about that space and what is contained there:
- What Vector knows is in the quad –a cliff, the edge of a line, an object with a marker symbol on it, or an object without a symbol (aka an obstacle),
- A list of what Vector doesn't know about quad – i.e. that he doesn't know whether or not there is a cliff or interesting line edge there,
- Whether Vector has visited the quad or not.

Vector subdivides quads to better represent the space. The quad probably is only slight bigger than the object in it. But the quad (probably) can be smaller than the object, to accommodate the object not oriented and aligned to fit quite perfectly in the quad.

How Vector map cliffs and edges

If a cliff (surface proximity) sensor has a large or significant change in value, Vector will make a note that there is a cliff sensor there. If the value has a smaller, but still noticeable change, he might make a note that there is a line edge there – an edge between a dark area and a light area.

How Vector tracks objects

Vector tracks objects (especially objects with markers) using the map, and other cross-referencing structures. Vector associates the following information with each object it tracks:
- The kind of object it sees (dock, cube, etc)
- A pose. The image twist of the marker symbol gives some relative orientation information about the object. Vector will uses this to compute an estimated orientation (relative to the coordinate system) of the object and Vector's own pose.

Vector can estimate the objects position from his own position, orientation, and the distance measured by the time of flight sensor.
- The size of the object. Vector is told the size of objects with the given symbol.
- The link to a control structure for the kind of object. For instance, accessory cubes can be linked and sensed.

If Vector sees a symbol, it uses the objects known size, the image scale, its pose (if known) and any time-of-flight information to; refine its estimated location on the map and to update the location and orientation of that object.

Path planning

Path planning is devising a path around obstacles without collision, to accomplish some goal, such as docking with the "home" (charger) or accessory cube. Intuitively, all you need to with a rectilinear grid is to figuring out the x-y points to go from point A to B. Vector (and Cozmo) is longer than they are wide – especially when carrying a cube. If this isn't taken into account by the planner, Vector could get stuck going down some path he can't fit in or turn around in.

Getting the Vector's lift and head position just right

The head and lift motors need to have their positions calibrated.

At startup, Vector performs a calibration procedure, "which is just an animation that pushes the head/lift to their hard stop." Both the lift and head have hard stops at their most downward position, which serves a well-defined starting point. When these motors reach the end of travel, the measured speed will fall below a threshold, and the software knows to zero estimated position. Vector's software has two backups in case the position is wrong. This can happen if the calibration was wrong – something, perhaps a block or impatient human companion – prevented the head or lift from moving further. Alternatively, if someone moved its lifts or head (since the position encoder is single step, Vector won't be able to tell which direction they were moved).

1. The body-board firmware has motor burnout prevent features. This quickly drops the power applied to the motor if there is a stall.

2. If a motor is stalled unexpectedly or the motor isn't stopped, Vector will schedule another calibration procedure.

Turn to record heading

The "TurnToRecordedHeading" is used to specify how Vector should turn to the previously recorded heading. The robot reach the target heading in the duration given, ramping up the movement speed smoothly until it reaches that position (within some tolerance limit).

Can Vector dance?

The dancing can be initiated two different ways. The first step is if a beat is detected. The second is if Vector is verbally told to dance.

If a beat is detected: A behavior node is regularly invoked as part of the behavior tree. This node has the pre-condition that a beat has been detected. This isn't quite the same as reacting sounds, but it is similar. If a beat has been heard, the "DanceToTheBeatCoordinator" proceeds in two phases. The first kicks off a helper behavior to listen to music. If it detects music (beats), it then fires off a dance behavior: there are two such behaviors, depending on whether or not it was on the charger. If there is no music detected – or Vector is no longer on his treads – this behavior exits.

If Vector is told to dance: A behavior node is regularly invoked as part of the behavior tree. Part of preconditions for being able to execute this node is that someone has given Vector a command to dance. This is done by the "respondToUsersIntents" condition including the "imperativedance" intent. If there is no such user intent pending, the node is skipped.

CHAPTER SEVEN

VECTOR'S ANIMATION

Vector animation is a scripted sequence of the ways Vector coordinates its movement, faces, lights and sounds. Animation is used by Vector to show an emotion or reaction to an event. The Vector's small body is what it leverages to make it carry itself quickly to depict an emotional reaction. Much of the Vector's animation is done by vic-anim. Animation is not possible without the animation files working synergistically to make it happens. There are about seven categories of these animation files, which include;

- JSON animation file: This describes how the backpack light should be coordinated.
- JSON animation file: This categorically describes how the cube light should be coordinated.
- Binary animation file: This coordinates sounds, eyes animation and the head and lift movement.
- Sprite sequences: These are PNG folder of images, which are programmed to display in sequence.
- Composite Screen: Displays icon and text information driven by the behavior and cloud server intent.

- Sound files: This holds pre-recorded sound effect.

Light Animation

The backpack lights are deployed to indicate the state of the Microphone, when Vector is charging, Wi-Fi mode and many other behaviors. The Companion cube's light does depict the process of setting up of the cube, entertainment and gaming. The light animation has three sources, which include;

- Binary animation file: This drives the backpack light
- The Cube's JSON file and backpack light animation: This JSON file contain files for the cube's light sequence, files that runs the backpack light sequence and two files working for animation trigger name.
- The Cube spinner game, which runs on JSON-driven light animation. This has not been enabled yet in any of the Anki's software. It was only planned to be released.

The backpack light illumination follows these sequences;
- The Cube Spinner name produces an animation trigger name

- The animation trigger name is mapped to an animation file
- The animation file provides the sequence to illuminate the backpack lights.

VECTOR'S VIDEO DISPLAY AND FACE

The Vector's liquid crystal display is used to show the Vector's face, accentuate its mood, form some forms of emotional connection and to show the result of its behavior. The Vector robot usually displays moving imagery on the screen. The drawn on screens are of these formats;
- Full screen sprite: This contains frames with PNG format, which covers the whole screen.
- Composited images and texts
- Procedural face, which Vector can use to draw the face in a complex way.

How Vector uses sound to convey emotion

Vector uses sound to display its emotion and activities, to talk, and to play sounds streamed from SDK applications and Alexa's remote servers. There are five sources of sound:

- Sound effects generated from playing pre-recorded audio files
- Sound effects obtained from parametrically generating audio
- Sound from an audio stream sent by the SDK application to Vector
- Text to speech
- A sound stream from Alexa Voice Services.

Vector has the ability to modify sound effects parametrically. The behavior-animation systems uses this convey Vector's emotional state – sadness, approval, etc. These are like intersectional tones that people make – grumbles, and grunting. When an action is being performed, the animation may trigger the sound effect, perhaps to "simulate the physical movement" that the Vector is making. The sound effect parameters are modified by the current emotional state, or anticipation – to convey whether he is struggling to do the task, is excited, is frightened, etc.

The sounds events for these are directed to a special game object just for them. Most of Vectors sounds are driven by the animation, and when they are sent to the audio engine, they are tagged with "Animation" as their game object. For the procedural sound effects, the events are tagged with "Procedural" as their game object.

Sound Command

The HTTPS SDK API includes commands that affect the sounds
- Audio stream commands
- Text to speech: An external application can direct Vector to speak using the Say Text command. The response(s) provide status of where Vector is in the speaking process.
- Vector's volume can be set as a setting using the "UpdateSettings" command and the "RobotSettingsConfig" structure or using the Master Volume command. Note: the volume levels using settings doesn't fully match those in the master volume command.

ABOUT AUTHOR

Derrick Grisham is a software expert who has spent a huge part of his life teaching and writing codes for software. He is a passionate instructor who finds joy in teaching other what he knows. He has an online community where he teaches people how to trouble-shoot and diagnose basic software issues.

Derrick holds an Msc degree in computer science from the University of Minnesota, USA. He is a lover of pets and admires their weird ways of relating with humans. He lives in Minnesota and is happily married with two beautiful daughters.

www.ingramcontent.com/pod-product-compliance
Lightning Source LLC
Chambersburg PA
CBHW070259220526
45465CB00004B/1668